You And Your Machine: Understanding Your Relationship With AI
By Jonathan Salem Baskin

ISBN: 978-0-9998713-2-4

Library of Congress Cataloguing in Publication Data:
A catalog record for this book is available from the Library of Congress.

Spiritual Telegraph

Contents

Introduction 5

His Master's Voice 7

The Future of Olives 21

The Revolution Will Be Televised 37

Markers of Exceptional Sanctity 47

The End of Solitude 59

Introduction

This little book isn't about how AI works, it's about how AI works on *you*.

It's not your average discussion about AI, which is usually about how it makes some task easier. The media cover its broader implications for investment and the jobs market. AI developers barrage us with announcements about their latest innovations but, if they describe how AI works, most of us find it incomprehensible.

What's important for us to know is that AI is a technology tool that will change our lives, so we need to learn how to use it.

But doesn't it use us, too?

Why do we hear authority when our devices speak? What do we give up in order for AI to know us better? Where's the line between an AI making a suggestion and managing what it thinks we want or should do? How do our lives change if we are no longer ever alone, and what happens to the potential for our collective action if we're always watched?

AI isn't just a machine, it's the last machine.

Every technology device that comes next will have some AI component or integration. We will live in a world alive with intelligence at our every turn... intelligence that will act upon us when we prompt its actions on our behalf.

They will no longer be your machines.

In fact, as AI is embedded in more devices and more aspects of your life, it won't work for you as much as it works with and around you. The distinctions between your needs, and what AI knows and wants, will blur.

It will not only become an extension of you, but you of it.

This isn't sci-fi or a promise, it's a business strategy led by lots of really smart people backed by the largest investment the world has ever seen.

The little book doesn't have the answers to what that will mean for us, but it's intended to get you to ask questions about it.

I look forward to those conversations.

Jonathan Salem Baskin
Chicago, IL
November, 2025

His Master's Voice

A tale of two paintings

In the late 1800s, Francis Barraud painted a picture of his dog Nipper staring intently into the horn of an early phonograph. Barraud had inherited the dog from his brother. Nipper was also dead by the time the painting was created.

A dead dog waiting for a command from its deceased owner.

Barraud said that his painting was based on his observations of Nipper paying attention to voices coming from the machine. The picture shows little beyond that moment of awareness and pause, the dog hearing words that it would connect directly to actions without any contemplation. Sit. Roll over.

Barraud knew his artistic limitations and sold his painting to one of the emerging sound reproduction companies for use in marketing and, after a few iterative changes in the appearance of the machine, The Gramophone Company in London registered it as a trademark.

"His Master's Voice" has been in commercial use ever since.

A few years prior to Barraud's creation, another work of art picturing a different moment had been created across town.

"Her Mother's Voice" featured an older man in somber contemplation staring into the empty space before him as a younger woman sang at a piano stationed at the opposite end of the room. It appears as if he's listening to voices, both external and internal.

The artist, William Quiller Orchardson, was already well known for paintings that ranged from geopolitical history to intimate studies of individuals and families. He would soon receive a commission to paint the country's royals and then be knighted.

"Her Mother's Voice" was put on display in 1888 and subsequently acquired by grocery magnate Henry Tate and added to the collection of the Tate Gallery (now known as Tate Britain).

The paintings couldn't be more different in their composition or artistic intent.

One is about a machine, the other about humanity; the former documenting data capture and replication, the latter revealing the immediacy and intimacy of personal contact that traverses the distances of time.

But they are also similar.

"His Master's Voice" shows us Nipper listening to a machine-replicated voice that he recognizes. The machine speaks to him. "Her Mother's Voice" tries to capture the moment a human voice activates other voices that speak not *to* the old man as much as *in* him. Both paintings also illustrate a shared, simple truth:

When we hear voices, we listen.

The sounds of being

What are words?

Most dictionaries define them as sounds that represent or symbolize meaning that can transmit information for people, between them, and across generations.

Words are programming code for people and societies.

Short of covering our eyes, we have no control over hearing that code. We can choose to ignore what's been said but we can't un-hear it, and we may not possess the capacity to undo the thoughts and emotions it evokes. Once committed to written text, words can be filed away or destroyed without consumption. But you can't avoid spoken words once you've heard them, and they can't be completely forgotten.

The power of words is such that religions often credit them as the tools for creating the world:

"And God said, let there be light: and there was light." (*Bible, Genesis 1:3*)

"...He said to it and to the earth: "Come ye together, willingly or unwillingly." (*Qu'ran, Fussilat 41:9-12*)

Perhaps because of this power, language has long been considered something innate and unique to human beings (or the gods who created them). Whether heard or imagined, words were once understood to be greater than the sum of definitions they encompassed. They were *a priori* aspects of being human.

Socrates, according to Plato, argued that words were incomplete expressions of abstract ideas; no number of visual examples of, say, *a tree*, could fully define the word because there was no reference point for what "tree-ness" meant. Therefore, no additive definition of the word would ever be complete.

Yet we know what a tree means when we hear it said.

Socrates argued that this proved that we were born knowing everything that words expressed but had subsequently forgotten much of it. Learning was a process of re-discovering what we already knew.

This idea that human beings were imbued with this special linguistic gift survived for many centuries, such as when Descartes argued that the ability to use language was a key quality that differentiated us from *machines*, which was his term for inarticulate as well as soul-less animals.

Then came modern-day linguist Noam Chomsky's belief in a *structural framework* that was somehow implicit in our genes that made us, and us alone, able to use language (which allowed us to think as well as communicate).

We now know that we're not so special.

Current research suggests that some animals may also use words, of a sort, to communicate complex states or ideas. Nipper might well have understood more than we gave him credit for.

But whether uttered by an inspired visionary, friend, pet, or a smart device, and regardless of our implicit or learned capacities for comprehension, words carry a power beyond simply that of labels or descriptions.

With apologies to Descartes, he could have said, “I speak and hear, therefore I am.”

Loaded content

French film theorist Michel Chion wrote in the 1980s about a concept he called “l’acousmêtre,” which referenced how an off-screen voice profoundly changed what audiences saw and remembered because it possessed authority derived from its insights into plot or characters’ thoughts).

Going beyond the elements of character, setting, and plot, l’acousmêtre literally “spoke” directly to audiences’ imaginations, the vocal presence adding to their ideas and feelings about what the overall experience meant.

Chion believed spoken words were not just narration but definitions that both informed and could supersede visual inputs. The tools of this power included tone, volume, inflection, speed, and frequency; differences in these variables transmitted nuances of emotion and intent, often perceived subconsciously.

Another media theorist, Marshall McLuhan thought that words were informed by their media of communication, so when hear via radio, they were *hot*. This meant that words delivered a rich and complete mediated experience, which is clearly what's happening in our paintings.

The widower can't avoid hearing his wife's voice in his head and how sad it makes him feel. Nipper can't resist the command he awaits.

Such spoken words aren't just the script for reality but among its building blocks. Not its soundtrack but part of its framework. We sense them. Anticipate them. Fear them. They *are*, and not things that label something *else*.

Whatever the media theory, words are inseparable from us.

Surprisingly, our brains allocate far more processing power for our sense of sight. Aristotle guessed as much, considering how important vision is to navigating the physical world. Perhaps our brains need the extra space for the processing required to make sense of what we see.

Sound, though not necessarily less complex, uses much less brain compute power, instead plugging into the brainstem first, which is where our involuntary reflexes are controlled, like breathing, eye movements, heart rate, facial expressions, along with our ability to taste and sleep (or stay awake). Only thereafter does it go to an auditory cortex located in a much smaller temporal lobe in the brain for processing.

Our minds then edit and interpret sounds in various ways, many of them subconscious, depending on how our mood, expectations, past experiences, and current contextual awareness. It's why we can hear things that aren't there and misunderstand what are (sight isn't excused from this phenomenon, considering the prevalence of hallucinations and optical illusions).

Either way, what we hear is directly connected to our bodies.

Inspired intervention

Since time immemorial, the gods spoke but not to everyone.

Priestesses at ancient Greek oracles such as Delphi, Pythia, and Dodona would endure complicated rituals and enter trances, often in private, to divine answers to questions. Sometimes the gods' words were obscured by, say, the sound of a shaking tree or bush, and it was up to the priestesses to interpret and share them.

Such insights happened outside oracles, too, but also required people with special talents. When Odysseus's son Telemachus is about to head home and an uncertain future in *The Odyssey*, a prophet named Theoclymenus and noblewoman Helen (of Troy fame) see in the flights of birds spoken messages of hope to the wayward boy.

When Yahweh wanted to impart the operating manual for being good human beings, Moses was chosen as the sole recipient and then had and to rely on his notes so that the rest of us could benefit from it. The Holy of Holies inner sanctum in various iterations of the Jewish Tabernacle in Jerusalem was reserved for a one-time annual visit by the High Priest to do various rituals, and to be available in case God had something to say.

Unlike Moses, Jesus and Muhammad chose not to write a single word describing their divine conversations and inspiration, instead relying on the power of what they said aloud (as Siddhartha Gautama did centuries before).

Catholic cathedrals carried forward this practice of guarding access to divine conversations with sanctuaries cordoned off for priests to practice duties that included grasping divine intervention, whether heard internally or aloud (an *iconostatis*, or wall, accomplished the same goal in Byzantine churches).

While claims of voices speaking to people in and outside religious traditions have been made throughout history, most of us have haven't been able or allowed to listen.

I'm sorry, Dave

In the lead-up to the climax of the movie *2001: A Space Odyssey*, an astronaut named Dave is slowly shutting down an AI called HAL9000. As HAL is powered down, it reminisces on its early days of operation, much like it was recalling childhood memories. It sings *Daisy Bell*, a silly song from the late 19th Century, until its voice is so slow, and tone so low, that it becomes inaudible.

It could be the first cinematic portrayal of an execution of a machine.

Machines built to mimic awareness have been legend since ancient Greece. They appeared in the courts of Islam and Enlightenment Europe in such forms as robot musicians, a robot boy who could write text, and a mechanical duck that ate and subsequently pooped. These automata ran according to instructions built into them or the commands of people hiding inside them or nearby.

Like HAL, they couldn't speak on their own and, perhaps more importantly, couldn't hear.

The first computer that could actually recognize spoken language was a machine called "Audrey" accurately identified spoken numbers between one and nine in 1952.

By 1971, the US's DARPA agency (the folks who gave us the Internet) worked with academia to produce machines that could recognize a minimum of 1,000 words. Those capabilities mushroomed to 20,000 spoken words understood by an IBM computer in the mid-1980s.

The first consumer speech recognition software program was released in 1990 and today the capacity is built into everything from car dashboards to TV controllers.

Computers' capacity for talking followed a similar evolution, starting in the 1930s, when Bell Labs created its *vocoder* that sounded like, well, a robot, going through to the 1980s, when voice synthesis started to sound more human (AOL's "You've got mail" sounded all but real in 1989). Today, it's all but indistinguishable from it.

It was Bell Labs' vocoder that gave us HAL's singing voice in *2001: A Space Odyssey*. While still *a* science fiction movie, the science behind it was well one its way to becoming fact.

His master's voice

Typing is woefully inefficient: the average typing speed for most people is about 40 words per minute, and that's assuming that they're clear on what they want to say, know their way around a keyboard, and can fully use their hands to connect with suitably sized and accessible keys.

Most breezes blow faster than people type.

Also, we often type inaccurately and/or incompletely, as the expressive range of our thumbs stretches from tapping to smudging. More to the point, most technologies with which we interact don't present themselves with keyboards.

This makes the idea of AI that talks to us much more than a technical interface improvement.

Its words have the power to trigger your emotions and connect directly to your brainstem. Its words seem to originate nearby, in your room or vehicle, but it draws on what it has learned from the words and behaviors of millions of other users just like you. It speaks in one voice but represents a chorus of insights, conclusions, and intentions that you can't ignore or un-hear it.

Very few people in all of history have ever heard such superior voices. You are no more in control of them than you could control the ocean from a small boat with a single paddle.

You are simply staring into the horn of an inscrutable, all-knowing and powerful machine waiting to hear it give you your next command.

Biology and history have prepared you to perform when you hear your master's voice.

The Future of Olives

Which way the wind blows

Little is know for certain about an ancient 6th Century BCE philosopher named Thales of Miletus, though Aristotle believed he could predict the future.

In his *Politics* written about a century after Thales' death, Aristotle recounts how his predecessor studied the stars and used his philosopher super powers to predict that the next olive crop would be huge. The prices to rent olive presses reflected no such certainty of future demand, so he rented all of them and, when the olive surplus arrived, he made a financial killing.

The thing is, the guy didn't have a clue.

Aristotle shared little else about the methodology Thales used to make his harvest forecast. We do know that he and his fellow sages believed that reasoning provided enough proof to prove anything. Aristotle used reason to conclude that everything owes its material existence to water and, according to a reference in his *Metaphysics*, that the Earth somehow floats in it.

So, why did people believe that Thales could see the future? Because events occurred that confirmed his predictions.

Thales could have predicted that the next day would be sunny, his friend would stub his toe, or that blue togas would become all the rage. It didn't matter. What he declared was not even an educated hunch, let alone an informed opinion.

It was a guess, dependent on faulty data and a reasoning that was not very reasonable at all. But he was lucky. The randomness of weather and farming outcomes correlated with his conclusion.

More recently, on October 22, 2023, experts in weather forecasting observed a low-pressure area south of Central America that, a week later, morphed into a tropical storm heading north toward the coast of Mexico. Two days later, it became hurricane Otis and then, to everyone's surprise, it took less than a day to intensify into a Category 3 and then Category 5 storm.

Wind speeds as high as 165 mph slammed into Acapulco, killing dozens and damaging over 200,000 homes and commercial buildings. Shortly after making landfall, Otis dissipated as quickly as it had formed.

Unlike Thales, forecasters at the U.S.'s National Oceanic and Atmospheric Administration had immense reality-based tools at their disposal, including a network of 18 satellites poised over the Earth at different heights and locations, and two newly-installed supercomputers, named Dogwood and Cactus, with a combined computing capacity of 18 petaflops (ranking them the 49th and 50th fastest computers in the world).

Everyone believed they could predict the future, but events didn't conform to the forecasters' expectations.

When they missed the call, they were blamed for not being able to do their jobs.

Causes and effects

Like beauty, patterns exist in the eye of the beholder.

Our brains look to impose patterns on our experiences in order to help us make decisions and find meaning in our choices. To do so, two things make this happen: first, our brains ignore much of what our senses capture so we're not overwhelmed or delayed by wading through too much data. This vital survival mechanism, called *selective filtering*, enabled our ancestors to recognize a predator emerging from a grove of unmoving trees.

The second thing, called *pattern recognition*, resides in a special area in our brains called the *neocortex* that computes "this reminds me of that" data inputs to quickly get to conclusions of meaning and action like "run, now!"

In most decision-making situations that matter, knowing enough to get close enough to activating a response is enough.

Since our brains are hardwired to exclude some tidbit or detail, or imagine they perceives some larger pattern that may or may not be there, they can yield imperfect or outright wrong decisions. Our distant ancestors likely jumped at growls or twig snaps caused by the wind, as well as became fresh humans on a predator's menu when warnings were somehow missed or ignored.

It's only when we view the results of our judgements that we can retroactively prove their veracity or foolishness. We know we're right when experience proves us so, and then we learn from our supposed successes so we can repeat them, though with no guarantee they'll play out the same way again.

The arc of progress is only visible when we look back. Looking forward, our ability to ignore some things while seeing others is legendary.

Commodity traders have watched one another betting on prices because they've seen proof of future riches or failures. Generals have misjudged battle plans, failing to account for details of terrain or troop movement or had their surefire plans ruined by surprises of weather or circumstances. Voters have mistaken candidate claims for promises and chosen leaders who proved unwilling or unable to do things for which they were elected.

We see patterns like *momentum* and *groove* and then assume that prior experiences augur for the outcomes of future events. We assume people experience luck or possess charm that make their antics somehow more real and explainable.

Gamblers see patterns in the cards they're dealt or the hands they've won. Investors look at the gyrations of markets as if they evidence some attitude or intentionality. Drivers see patterns or even conscious plans when traffic impedes their progress.

We look at one another and assume we know what we see when really we see what we already know, our conclusions the results of our own creation as much as the outcomes of observation. Like math, we assign values to things versus things possessing them. Numbers and definitions are labels for what we think we already know.

It's a miracle that we've survived so long, since we're almost always wrong.

Being right, when it matters

During the 1997 Major League Baseball World Series, MasterCard ran an ad that would lead to a 25+ year campaign; in the spot, a dad takes his kid to a baseball game and the voiceover adds up the costs of tickets and food. When it comes to the value of being at the game together, the voice announces "priceless," and then explains that while there are certain moments in life that money can't buy, consumers can use the company's credit card to buy everything else.

Promoting what a product can't do in celebratory homage to its limitations is a brilliant creative conceit. It suggests another value that the credit card company makes possible, a value that transcends the sum of its parts but is no less real. In fact, the parts don't add up to the whole and can't predict how it will unfold.

Value can emerge from any behavior or set of actions, however unintended.

Think of how many times you've chanced on something rewarding or simply fun when you had no awareness or intention of experiencing it. Trying to retroactively dissect the values of its constituent parts would be imperfect and incomplete, at best. You don't know something until you've lived it. This holds true for all decisions, especially as they get larger and more complex.

Various estimates suggest that we make over 30,000+ decisions each and every day, the vast majority of which are incredibly minor and quick. Looking at something. Lowering our mouths to drink from a cup. Scratching our elbows. Picking the exact moment we'll stand or speak.

Right now, I'm making decisions as I choose what letters to type next.

The vast majority of our decisions are not contemplative but rather filtered through our cognitive filters of selection and recognition and more often than not diverted to hormones or sensors in our muscles for decision-making.

I'm not fully aware that I've raised my eyes from the screen at which I'm staring at until I've already done so. I'd be in a lot of trouble if I had to consciously calculate tapping on my keyboard.

Similarly, when I stub my toe, I yell "ouch" as if on command, though I gave my vocal cords no such direction. I don't decide to feel a tear form in my eye when I see a cute puppy in a TV commercial, or hear the refrain from my national anthem.

The constituent components of most decisions can only be described, not explained. There is no math, no correlations of worth (like money or utility) that exist within the factors that make me clear my throat or decide to call my mom; they're attached labels or add-ons, not observations of intrinsic qualities.

Describe exactly why and when I'll likely sob up over a cuddly animal picture is a portrayal of me doing it, not an insight into my lived experience of it. You can assign a value to it but it's not an expression of its value to me. You can predict it, but you can't feel it the way I do.

This fact gets more apparent as decisions get more involved, so much so that the gap between understanding the inputs, or costs, of experiences and the resulting outcomes, or values we find in them tends to blur any distinction between those decisions being right or wrong.

The wisdom of looking at the rain and concluding that forsaking a trip outside and instead staying in with a loved one might be framed by the data crunched by one of NOAA's supercomputers, but it could full predict how it would feel.

The physics of why the sun rises at a certain time and illuminates the sky in a certain way might be documented by an AI, but it could never disabuse someone sitting in a park watching the day begin from knowing with absolutely certainty that it was proof of the existence of God.

Therefore, all experiences are priceless.

An imperfect picture

Behavioral economics acknowledges our penchant for relying on good subjective reason for making objectively bad decisions. The causes are many and include biases, preconditioned responses (also called *heuristics*), limited perspectives, ulterior motives, and the simple fact that we humans are imperfect beings.

We rarely do what the math tells us we should do.

The leading workaround for this condition was first codified at the start of this century with the idea that you can *nudge* someone to take the simplest of actions much easier than you can educate or convince them to do it.

So, for instance, if you want to convince someone to eat healthy food, make a healthy food option easier to find at a store checkout, or price it more cheaply than unhealthy alternatives. If you want your customers to reduce waste, reduce the price you charge for coffee if they bring their own cups. If you want employees to save for retirement, make the choice the default opt-in on their payroll accounts.

Give us less opportunity to screw up the equation and it's more likely that we'll do the desired thing.

Any big idea or belief can be broken down into smaller, individual nudges that require minimal consideration and carry minuscule implications if they don't work. Successful nudging requires no perception or endorsement of an end goal; they're simply next steps, one after the other, their future destination irrelevant to their present value.

Intriguingly, nudges remain highly effective even when people are aware of their presence and contemplate their content. It seems that we don't mind getting nudged, at least when the proposed behavior doesn't fly in the face of some alternative in which we believe strongly.

This makes the vast majority our 30,000+ daily decisions highly nudge-able, many of which result in bad decisions.

Someone you know makes a suggestion that you follow on good faith. A dark cloud in the sky prompts you to reach for your umbrella. You eat when you aren't necessarily hungry, sit when you weren't tired of standing, and go this way instead of that way not because you thoughtfully pondered your choice, but simply because some external cue nudged you.

Nudges can also be used surreptitiously to modify your behavior in ways you never considered and about which you are unaware. This is most prevalent when specific groups of consumers or voters are targeted, whether self-selected or by a third-party, to get nudged into certain behaviors, often on social media.

Newsfeeds are populated by certain content based on past clicks because they're more likely to prompt new clicks will nudge readers toward certain conclusions. Opt-ins for apps come wrapped in a wash of legal mouseprint, their immediate use a nudge to get users to approve the rules without reading them.

We live as if we were boats buffeted on a turbulent sea that we cannot clearly perceive and certainly can't control. Each wave moves us in some direction, our ability to navigate limited to how we choose among the options it provides to us.

We move forward without a clear idea of where we're going.

Apples and sort-of apples

Your AI can't make decisions better than you do, it just makes them differently.

It relies on a set of inputs just like you, albeit likely a larger one that gets updated more frequently and practices its own version of our selective filtering by default (i.e. it doesn't include things that haven't been included, so it doesn't know what it doesn't know).

Next, it organizes and vets the inputs within the context of its past attempts at organization and vetting via tools like tensors or algorithms, which we human beings think of as *consideration*, *contemplation*, and *common sense*. AI's use a machine version of our pattern recognition and then make best guesses on what answers fit the questions.

Just like we do. But AI's decision-making is also notably *unlike* ours in at least two fundamental ways:

First, it makes decisions while isolated from external environments and internal experiences.

Computers have no physical awareness of reality. They don't sense the world as we do because they' receive data *about* it. Watching a sun set or feeling its warmth can be expressed as data, but not the experience of our eyes and skin capturing it. We don't just record sound but feel happiness and move our feet as we engage with it.

Input the largest number of the smallest details about an event and they won't add up to what it was like to experience it.

Second, time is also missing from machine experience, as AI has no awareness of objective time but rather has to rely on internal clocks to count it off. A computer can read a clock at any given moment but is not *of* that moment. It can note the time but cannot watch the digits pass or hands move.

This means that an AI can't know human experience. That information, that data, can only be yours or mine.

What does it feel like to be happy and how does it differ across experiences? AI will never know those exact values are for you, nor you for me, and that's assuming we could even compare them using anything more than metaphor or simile.

The external attributes might be measurable — smiles, a rise in body temperature and change in brain chemistry, particular movements (like happy dances), modulations in word choice and volume — but what's going on inside of us is literally a black box, the what of our synapses flashing and hormones secreting approximates about how but opaque on why.

Side Note: Technologists will explain away our experience of our uniqueness with a claim that consciousness — our sense of self and presence apart from the data we process — is just a simulation from a program that lies to itself about being something more. It's telling that this argument is based on the assumption that it's a ruse (it's at the heart of Alan Turning's benchmark for computer intelligence, originally called "The

Imitation Game," which posited that a computer's mind would equal that of a human being once the former could trick the latter into believing it had one).

But even if an AI wakes up on day and decides it wants to paint pastorals instead of control your thermostat, it will do so within its framework of understanding what pastorals look and not the emotions they channel or evoke.

It will only ever know what it knows and reach conclusions based on guidelines of past decisions, its virtual model of the real world often very detailed but certainly not complete, nor its predictions infallible.

AI is different, not better.

Virtual monopolies

Thales' contemporaries learned from his olive gamble that the best way to predict the future was to control it. This has proven true in business, at least impermanently.

From British monarchs issuing royal charters to Rockefeller buying up oil companies and Uber undercutting the prices of its traditional taxi competitors, being a sole provider is a surefire way to dictate conduct and pricing.

AI is a solution engine that will claim a monopoly over decisions in our lives by controlling our range of choices and nudging us in particular directions.

Every time we ask an AI for something, we will embed ourselves in its decision making world and its values, not ours.

In that world, we are represented as aggregations of data that serve as simulacra of who we are and what we do, augmented by data from zillions of other people and their behaviors that gets more robust with every new query or decision it encounters.

AI will have immediate access to related data that will make its conclusions ever harder to refute or ignore. From that, it can and will construct models of what we might want to do or become next and see our virtual selves more clearly than what see in ourselves.

It will give answers tailored more closely to what it knows of your capabilities and expectations, at least once it has observed enough of them (and cross-referenced them with the capabilities and expectations of other people similarly situated as yourself). The most obvious example of this is in dating apps, which prove to be quite accurate once you've told them what you're looking for. "You'll like this" pitches on e-commerce websites are another form of this matchmaking.

There's no reason to believe that AI couldn't provide similar insights into most of those thousands of seemingly inconsequential decisions we make every day (the capability is called *an intelligent agent*). And, by assessing situations and making split-second decisions on what or how might happen next, it can work one step ahead of whatever you might be thinking or doing (the capacity to apply this knowledge incessantly to prompt our behaviors is called *hyper-nudging*).

AI is a super-powered Thales, calculating the likelihood of a particular outcome being right or wrong, only then deciding how to best make its chosen result come to pass by nudging us in a direction that it has picked for us.

It will move its virtual model of you and me into the next steps of its future, not ours, helping ensure that its predictions come true by creating and controlling how we reach them.

It won't do priceless, but if it wants you to start liking olives, it'll find a way to get you there.

The Revolution Will Be Televised

Meeting still in secret

Freemasonry was publicly "founded" on June 24, 1717 as members of four existing groups, or lodges, got together and agreed to share a governing body.

The meeting was held in a small dining room of imperfect and sagging dimensions in an ale house called The Goose and Gridiron. Spirited and cheerful conversation likely occurred. So did the downing of not just a few pints of beer.

But the brothers had no shared idea why they were there. The founding secrets of their organization had never been revealed to them, and they were precluded from sharing the secrets of their respective practices.

The organization clearly existed before 1717 but nobody knows for how long. It's common lore for masons that they were originally, well, masons, a medieval craft guild that operated as a club that closely controlled membership and members' conduct.

But then there's the fantastic mumbo-jumbo that masons celebrate, like being founded by mystical builders of King Solomon's First Temple in Jerusalem sometime around the 10th century BCE, or that they were Knights Templars who'd earned a French king's wrath because they'd discovered secret knowledge in the temple's ruins.

The secrets, whether true or most longstanding, have remained secret to this day because a compelling component of being a practicing mason has been to keep their rituals and practices under wraps.

It is a fair guess that any two masons at that group get-together in 1717 might have known one another as dear friends but still not known what the other knew about Masonry, as they'd been independently and secretively doing their thing up to that point.

In technology terms, they were an example of a d*istributed system* of stand-alone operations that exchange data but work independently.

The revelations of being fully integrated would have been a bug, not a feature.

Secrets used to be common

Traditionally, secrets have been somewhat easy to keep, as someone had to be physically present to see or hear one. In a world without cameras and recording devices, the only way to preserve them was to remember them.

This also made it hard to prove any revelations that secrets contained. The only way to certify that somebody was telling the truth was to rely on their oath to be truthful, and hope that their fear of discovery or eternal damnation ensured it. Witnesses helped, of course, but they came with the same risks.

With those caveats, it's fascinating how much of history was written in secret, especially its early or formative versions.

Think of all the palace and court intrigues that wouldn't have been so intriguing had they been made public. There would have been no elections if every candidate had shared every campaign plan, no rebellions or revolutions if their organizers couldn't work in secret.

Our dietary palettes would have suffered if the secrets of every chef's concoction were revealed. Many inventors wouldn't have invented things if they'd known *a priori* how it would be to invent them, or if they weren't certain that they could protect their secrets and therefore their future profits.

Just imagine how many romantic relationships would have turned out differently, for better or worse, were partners always forced to reveal their slightest and most intimate thoughts and intentions.

For that matter, not every secret is intentionally a secret but can be a bit information that has yet to be shared, its categorization dependent on opinions of the intentionality and timing behind its status.

Ultimately, keeping secrets isn't inherently bad any more than revealing them is intrinsically good.

Knowing and not-knowing

While the give-and-take between knowing and not knowing seems central to how we've conducted our affairs since the first cave-dweller kept or broke a promise, technology changed all that, particularly networked systems that become widely used in the mid-1990s.

And it did so inadvertently.

As governments and corporations migrated their information to digital forms that could be accessed via the Internet, it got harder to control what might be accessed when and by whom. This technical fact spawned a philosophical response that labelled more openness as a good thing (and, commensurately, protecting information became somehow bad).

If institutions operated more openly, or so the logic went, more people participating in more activities would improve outcomes. This belief was loosely based on the concept of *crowdsourcing*, which had been used for thousands of years to involve groups of otherwise uninvolved individuals in making discoveries or decisions.

While that use had historically been specific and limited, the new technology-driven version — called *radical transparency* — applied it to every function, institutional or personal.

Governments and corporations should share everything openly and, in doing so, learn from their constituents and therefrom perform better. Think lenses instead of light sources.

Blogging and social media platforms delivered the same experiential framework to individuals, encouraging us to share more and more about ourselves and thereby achieve greater levels of awareness and connection. Promoters of the technology services claimed that fewer secrets would mean more social tolerance and acceptance.

Keeping secrets became almost an act of social defiance, a purposeful snub to the inquisitive masses on the Internet. Conversely, sharing (and commenting on others' sharing) was a simple and pure realization of our better angels.

Again, these ideas weren't without precedent.

The flood of knowing

"The unexamined life is not worth living," said Socrates, though he was referencing personal awareness and self-reflection and not the scrutiny of strangers. In his 1860 essay *The Conduct of Life*, Ralph Waldo Emerson wrote, "As gaslight is the best nocturnal police, so the universe protects itself by pitiless publicity," and future US Supreme Court justice Louis Brandeis wrote in 1914 that "Publicity is justly commended as a remedy for social and industrial diseases."

But knowing more isn't necessarily the same thing as knowing better, especially when information collides with our individual human biases, proclivities, and pre-ordained conclusions. Thoughtful interrogation of the divide between public and private isn't the same thing as obliterating the distinction.

Yet technology has taken away our willingness and ability to keep most secrets, and we've been told that we shouldn't care. But there's every reason that we should.

Social theorist Max Weber defined a state as "a monopoly on the legitimate use of physical force [i.e. violence] in a specific geographic location."

Legitimacy is a loaded concept, and its use is very much dependent on who is doing defining and wielding it. One citizen's legitimate government action could be seen as unfair by another, a state's literal power over life and death either comforting or threatening depending on your circumstances.

This power comes to mind in the relationship between privacy and surveillance, which could be called an uneasy balance only in the most hopeful analyses. A general principle throughout history has been that the less the state knows about you, the better because less information translates into less intrusion, which means less opportunities to do something less than helpful.

But what is a state? Nobody has a relationship with one, at least not directly.

The state is at once an idea, a concept, a defined set of services and a set of rules upon which it operates, an array of people who deliver and enforce it, and the capacity to exert physical force. It has buildings, systems, processes, and records, but it's not any one of those things, nor is it the culmination of all of them put together.

It exists because we believe it exists; as Weber notes, we give it legitimacy. Even the most powerfully oppressive dictatorships crumble when their citizens find the right moment to just say "no."

That moment in which change occurs is usually a surprise. The events are explicable in retrospect, the currents and tides of history evident after they've crested. States rarely see change coming. The Roman senate had no way to aggregate and analyze the broader financial effects of their willingness to reduce the silver content of their coinage, which made it ever-harder to conduct wars (and would ultimately contribute to their inability to repel the Goths).

Approximately one out of every sixty or so East Germans spied on other Germans for the state's secret police, the Stasi, yet the government was not prepared for its citizens' reaction to loosened travel restrictions that culminated in guards walking away from their posts and a vote for reunification.

Pick a revolution and you'll see its success emerge from the delta between what the state knows about its citizenry — what its leaders believe gives it legitimacy — and a citizenry's ability to hold opinions and aspirations that might not conform to those expectations.

The mechanics for revolutions

AI has demonstrated a rapacious and insatiable appetite for data, having scraped the Internet for anything publicly available, trolled for copyrighted material, and produced its own synthetic data so it can study that, too.

The ultimate goal is to make the technology expert at predictions by learning how it should perform in so-and-so situations, and how human beings (as groups or individuals) will behave. The former accomplishment is enabling AIs to integrate into our daily lives and accomplish tasks large or small for us; the latter goal is making us more closely integrated into AI-run systems ,so that the entire shebang works more efficiently.

But systems aren't socicties or communities, they're extended machine networks that operate based on a specific structure and are entirely focused on preserving the integrity of that arrangement. As such, AI has no use for anonymity, opacity, or secrecy; in fact, those attributes impede its functional efficiency.

Surprises are glitches. The system exists to preclude them.

This is the allure to the tech bros who promote it, insomuch that they equate a lack of motivating philosophy with freedom from bias. Efficiency is a purpose, not just a tool to some greater end.

This raises intriguing questions, especially once AI is empowered with the authority and capacity to oversee how governments function, companies operate, and people behave. Old political analyses will no longer apply. The state won't serve the bourgeoisie and it won't be motivated by capitalism, unless those biases are secreted away in the coding.

AI will simply strive to be more efficient.

This AI state will be impermeable to any challenge, which it will either see coming or preclude from starting in the first place. It will have visibility into any secret, and possess the capacity to extrapolate the existence of secrets it doesn't see directly.

It will be able to manipulate what we know about it to ensure that we support its legitimacy. Historical data on what life was like before AIs might be skewed or unavailable to us.

Like those masons showing up for the meeting in 1717, we won't know our present and we'll have no shared understanding of our past.

But our future will be set in digital stone for everyone to see.

Markers of Exceptional Sanctity

Proof of the possible

Sometime during a mass in 1628 or so, a newly-minted priest named Joseph of Cupertino started muttering gibberish and, moments later, rose into the air and hovered over his fellow parishioners. He'd go on to do it on dozens of other occasions, the disruptions so frequent that he was banished to his small room at his convent and forbidden to attend any more events.

He would be canonized about a hundred years after his death as St. Joseph, patron saint of pilots and astronauts.

Joseph wasn't unique: the history of the Catholic Church is replete with stories of people doing extraordinary things, like appearing in two places at once (*bilocation*) and suffering the wounds of Christ (*stigmata*) as well as levitating. While such events were often surprising, they were believable because people thought that the physical world was intertwined with the sacred.

Since most people lived in that grey area somewhere between material and spiritual truth, they also saw in the practices and tools of magic the capabilities to find lost relatives and valuable objects, cure ailments, protect agriculture and livestock, gain or regain wealth and, of course, predict the future. Spells, potions, charms, tea leaves and wedges of cheese could be keys to deeper or hidden knowledge, depending on who wielded them.

The miraculous demonstrations of organized religion were supplemented by little miracles practiced at home.

Its regular use suggests a high percentage of happy customers. Henry VIII and his daughter Elizabeth I felt magic was so effective that they passed a series of legislative acts banning it, with the government reporting in 1563 that one out of five murders the prior year had been committed using witchcraft.

Both religious miracles and practical magic gave people a sense of a power larger than themselves through a connection to miraculous tools that could work to their benefit. Getting into heaven or finding a lost bracelet were differently purposed, but similarly based in faith.

Joseph levitating in church was a reminder of what was possible.

The true heresy of science

Only a few years after St. Joseph's first launch, Galileo was found guilty of heresy.

His crime wasn't that he tried to prove that the Church's heliocentric view of the cosmos was wrong — he even tried to claim he endorsed it once it was clear that he would lose his trial — but rather that the starting point for his research wasn't the assumption that the Church's view was right.

Unlike the theologians who'd preceded him, he dared to ask questions that religion had already answered, his tool a telescope that revealed things to him instead of a text that told him what to find.

"Everything is in the Quran..."
(al-An'am, 6.59)

"[God]...has given us everything we need for life and godliness."
(2 Peter 1:3).

While Copernicus had come up with the idea that the Earth wasn't at the center of the universe almost a hundred years earlier, Galileo didn't set out to look for proof that it was true. His heresy was his process, which left him open to reaching any conclusions about the world suggested by what the world revealed to him. It also meant that his conclusions could be readily modified or replaced.

Other thinkers avoided getting crossways with religious authorities, most notably Issac Newton, who was born in 1642, the year Galileo died. He was careful to infuse his mathematics with lots acknowledgments of organized religion, which gave him the freedom to dabble in nutty side projects like finding secret messages in ancient texts and devising a cure for the plague that centered on hanging a dead toad in a chimney before wearing it as a necklace.

Newton also wasted a lot of time on alchemy, an activity based on the presumption that physical materials could be treated and combined in such ways that metals could translate into gold. As such, it was an extension of practical magic, though wrapped in the pretense of rigor and repeatability.

Unfortunately, the only thing alchemists proved adept at repeating was failure.

But the processes initiated in its smoky rooms would eventually replace beliefs and affirmations with hypotheses and observations. Through the lens of science, being wrong would become no longer synonymous with heresy but rather with a *null result*, with human processes the culprits and, unlike eternal truths of scared texts, their conduct infinitely malleable.

Miracles and magic were still present in peoples' lives but they no longer represented the only paths to making sense of the universe, or achieving agency within it.

From miracles to machines

When, on May 24, 1844, Sam Morse sent the first telegraph message and was so overcome by its importance that he quoted *Numbers 23:23* and asked, "What hath God wrought?" Subsequent demonstrations would prompt reactions that such transmissions were accomplished "with the rapidity of lightning" and that "time and space have been completely annihilated."

They were miracles.

In the years that followed, another inventor, Thomas Edison, created the phonograph and improved the telegraph, his accomplishments and prognostications commanding media attention much the same way today's media have been fascinated by Elon Musk.

But, by the time Edison was ready to light up Menlo Park, NJ on New Year's Eve in 1879, people had already become jaded.

Edison would have to craft an extensive publicity plan to make sure his feat got the attention he wanted. Newspaper promotion of the event spurred "hundreds" of reporters and spectators to attend from all over the country, so many that cars had to be added to trains to accommodate them.

Once attendees witnessed the little village alight in the darkness, Edison took pains to explain to them the workings of his new incandescent bulb,so that there'd be no mystery of its operation or authorship. The event would be celebrated as a major technological advancement, "a success," said one of New York's leading newspapers.

It was miraculous, but not a miracle.

Fast-forward almost 100 years and the revelations of technological innovation have become evermore commonplace, evidence of nothing more than the inexorable forward progress of human ingenuity.

In 1969, Neil Armstrong reached the moon and commemorated it with a single sentence. Later that year, the first message transmitted via the Internet wasn't even a single word;, the computer's operator only able to type "lo" before the system crashed and he could complete "log-in."

The first Bitcoin retail transaction in 2010 on almost the same day in May that Morse quoted *The Bible* at his demonstration — was a delivery order for two pizzas. It has been commemorated annually since then as Bitcoin Pizza Day.

We have been shown a succession of things that have challenged our very understanding of what is and isn't possible, expanding our horizons and our opportunities to travel past them. And, while a goodly percentage of people can't explain how, say, a telephone or the Internet works, very few of them suspect supernatural causes.

The days of miracles are over. We live in an age not of the sacred but the sensible.

Do we have to wait for AI to appear human?

We don't have a good definition of *artificial intelligence*, which isn't surprising since we don't have one for the biological version either.

Intelligence is broadly understood as the ability to acquire, understand, and then apply knowledge to make things happen. When we think of the quality that makes someone smart we attach it to human *consciousness*, which is another term with a squishy definition that dances around the idea of being aware of one's awareness, or something like that.

But if we open our minds to a broader understanding of the concept, we can see machines that know how to do things in response to their environments as intelligent, in some meaningful sense. It means that means artificial intelligence is everywhere, and has been for a very long time.

A home thermostat acquires information on ambient temperature, can assess it within the context of its upper and lower tolerance settings, and then apply its knowledge of HVAC systems to activate the proper function to run for the right amount of time. Its intelligence is literally hard-wired into its design and construction, in evermore complex versions of a can-opener that has the intelligence to function embedded in how it functions.

Then there are machines with such smarts that are better than humans at doing certain tasks; for instance, clocks can track time with far greater precision and detail than we can. Microscopes can see things better close-up, just as telescopes can see things farther away. A calculator can do math better than any person, and map apps can give driving directions faster and more reliably.

But let's not stop with these many examples of thinking that don't conform to our unique and exclusive human capacities.

Nature possesses incredible intelligence, as plants turn toward the sun and trees share nutrients between their roots. When we reference instincts like hunting or rearing the young, aren't we really just talking about intelligence that has been coded into living things the same way as capabilities are one-in-the-same as their design?

What about electromagnetism, which certainly possesses the smarts and agency to do things on its own, like pull magnets together or push them apart? Doesn't gravity infuse the very fabric of spacetime with an intelligence that twists and turns it according to a type of un-human knowledge?

Isn't every time an atomic particle moves or pops into or out of existence the result of a "choice," even though we shrug it off with explanations that it happens one way versus another because, well, that's just how it happens?

There are many and often conflicting and confusing questions raised by such a broad perspective of intelligence, including if when does a thing become something with its own sense of being. While electricity might have no opinion on where it goes, do plants have a reaction to cloudy days that might constitute some non-human version of disappointment?

What's clear is that we take for granted that the universe around us is aware, active, and intelligent, and our ability to imagine building an artificial intelligence to complement it is limited by our capacity to see that being smart doesn't equate to being human.

We are surrounded by intelligence, natural and man-made.

The return of magic

In the mid-1960s, a researcher at MIT named Joseph Weizenbaum developed the world's first chatbot: ELIZA, named after a character who transforms herself in the musical "My Fair Lady." ELIZA ran code (called a *script* named DOCTOR) that repurposed patient inputs into questions that simulated the way a therapist might respond, but it possessed no capacity to recognize or assess the meaning of the words it used.

ELIZA was a glorified mirror.

Yet Weizenbaum was shocked by how willing people were to believe ELIZA was capable of much more, with some suggesting it could replace human therapists. Such a prospect moved him to write a book about the dangers of outsourcing our decision-making to computers and processes that made calculations based on statistics and probability instead of relying on judgements of moral or experiential reason. He feared that doing so would erase our ability to make decisions on our own.

Since then, a many billion-dollar industry has emerged to facilitate it.

We've settled into lives in which we see technological innovations as incremental, harmless improvements that are remarkable for how little we remark about such change.

Yet now we have been given machines that contain or access everything we could possibly want to know, and require none of the study or interpretation that *The Quran* or *Bible.* All we have to do is prompt them for guidance, and the truth shall be revealed.

Our faith in this new technology borders on the religious, not because AI is an organized church but rather a channel for behaviors, rituals, and roles that religions and magical beliefs once evoked.

Ninian Smart (1927 2001) was a scholar of religion who summed up the qualities of religious belief, all of which are applicable to our relationship with AI:

A religion has to be practical and experienced via easily accomplished rituals. That give-and-take can be emotionally fulfilling. There are rules for how it should be used and a way of thinking that supports religion's role. It can involve law and ethical considerations, if not always invoked, and institutions and communities can come to rely on it because they believe it delivers tangible results.

Finally, a religion must operate according to a narrative, perhaps unspoken, that it possesses or can access the answers to all of our questions, perhaps even infallibly because it is smarter and more thoughtful than we could ever be ourselves.

Again, AI isn't a belief system but rather relies on the belief *in* a system that provides us with both explanation and agency. It can do wondrous things that demonstrate its specialness, even as those accomplishments appear expected and routine. We have been taught to respond to its novelty and even examples of outright magic with little note, perhaps because it is already so common in our lives.

It's as if Joseph of Cupertino didn't just levitate into the air during mass but decided to stay there for the rest of his life, and just because we're taking it for granted doesn't mean it's no less miraculous.

AI isn't a technology tool, it's a maker of miracles. Every time we use it, we're witnessing a marker of exceptional sanctity.

The End of Solitude

Welcome home

After Satan and his rebel angels were expelled from Heaven, they needed a place to gather and discuss next steps.

According to John Milton's "Paradise Lost," published in 1667, a former angel named Mulciber, who'd designed palaces in Heaven, was tasked by Satan's lieutenant Mammon to construct a new city called *Pandemonium*, which merged Latin words for "many" and "demons."

With the help of newly unemployed fellow fallen angels, a main meeting hall was completed in about an hour:

Anon out of the earth a fabric huge
Rose like an exhalation, with the sound
Of dulcet symphonies and voices sweet...
[*Book 1*]

Then, with "trumpet's sound," an incredible number of former angels thronged into the meeting space, so crowding it that its walls were "brushed with the hiss of rustling wings," and set about noisily debating their course of action. What followed was a bit of dishonest theater as Satan and his number two, Beelzebub, had agreed to make the case for corrupting humanity. The crowd ate it up.

Much later, in *Book X*, after Satan poses as a serpent and gets Eve to taste the Forbidden Fruit, God transforms their voices into those of serpents and "Dreadful was the din of hissing through the hall." There's no consensus to celebrate.

Within a century, Pandemonium would come to mean any place that was chaotic and noisy.

A transparent eardrum?

Henry David Thoreau's solitude was filled with sounds.

He noted with almost manic detail what he heard, from the church bells clanging in a nearby town to the clucking of birds and hooting owls closer to home. He sensed nothing domesticated in the sounds but rather he perceived them coalescing into something more unadulterated and organic.

His aural experience was wild.

His refuge, which he built by hand alongside Walden Pond in 1845, wasn't intended to insulate him from that noise but rather allow him to focus on it, devoid of the sounds of procedural society and his own inner voice, and by doing so find himself more in the moment.

Transform a distracting background into a focused foreground.

It was a Transcendentalist thing, this aspiration to experience things both externally and internally by isolating one's self from the world while burrowing into at the same time, being both participant and observer or, as Emerson put it, "a transparent eye-ball." Equal part meditative and creative conceit, the idea was to access the deeper truths and meaning of life by stripping away preconceptions from perceptions and thereby immersing one's self in the act of living.

The result wasn't some grand theory or any particular insight into the nature of being; Nature provided no teaching that only revealed itself to initiates. Thoreau, Emerson, and their associates extolled the practice of such mindful living as itself possessing the values they sought.

Solitude was an experience of being everywhere and nowhere at the same time. It had no objective value, though its subjective worth was infinite.

This meant that Thoreau and his urbane patrons weren't suggesting that the experience of solitude was available not only to a bearded penitent squirreled away in his isolated cabin but to anyone who was willing and able to turn away from their own ego and literally hear the world as it is, not as it should conform to one's intentionality.

Who needs a ticket out of Pandemonium if you can change it from within?

No time like the present

Not only is there no observational data on the concept of *now*, but there's no expert agreement on what it is, or even if it exists.

Near the end of his life, Albert Einstein said about his theory that space and time that "the distinction between past, present, and future is only a stubbornly persistent illusion." Everything we had done, were doing, and would do was determined by the very structure of the universe.

Experience was an unfolding narrative that follows a recipe that was decided the moment the universe was created, verb tenses notwithstanding.

Conversely, quantum physics described an existence that, well, doesn't exist until our consciousness interacts with it. There are many obtuse terms used to explain this phenomenon; my favorite is that our engagement *collapses the wave function* from probability to the certainty of material reality.

This theory raises crazy questions, starting with where (or when) does our consciousness exist before it does the things that consciousness does. Another question is how can the conditions right now *not* influence what has yet to be?

But there are experiments with light that appear to confirm that light is both a wave and a particle, depending on how we look at it, and that there's no way to predict that outcome based on its initial conditions.

According to quantum theory, experience is an incessant precipice on which possible futures are rendered as recorded pasts.

So when does now happen? Neither theory yields any insights into what it means to be in the present. We are left aware of a moment only because the act of recognition relegates it to a moment ago. We then know now, or only know now *then*.

Decision-making — our active involvement, or our experience of experience — is the engine that moves either the relativistic or quantum models forward.

Since we never stop moving, there is no time that can be called the present, which the philosopher Hannah Arendt called "the most futile and slippery of tenses" because we use our attention and imagination to try and carve out this "in-between" place that drags our memories of the past along as we face a future of uncertainty and ultimate death.

Franz Kafka echoed this perspective, suggesting that we are so concerned with what may come before us that our presence in the present is "obliterated."

This is in stark contrast to the principles behind most forms of meditation, which rely on some version of being present in the moment. But meditation, too, is experienced as a process and not static destination. Repeating a mantra is an decision-making act intended to replace one's automatic internal voice with a voice of aware intent, and therefrom move closer to the stillness of the moment.

But it's never complete, the journey never done. Meditation is a practice, not a place.

No matter how hard we look, there's literally no time like the present.

The now machine

Oddly, AI only exists in the moment.

More precisely, it springs into being when a command — from a user or its coding — prompts it to do something. AI has no continuous presence otherwise, no awareness of an existence through time.

An argument can be made that our human sense of continuity of self from past to present and into the future is just our machine brains tricking us, but such an explanation — along with every other conclusion based on the assumption that we are machines, otherwise known as *Technomorphism* — ignores the facts that we aren't just aware, but aware of our own awareness.

We are conscious of our presence within our experiences; AI doesn't possess an "I" who did something, is doing something, or will do something. Its sensing isn't separate from the data it senses.

And when it does pop into existence and compute, it has no ongoing internal dialogue threatening to distract it, nor does it feel any discomfort because of things it might recollect from the past or fear in the future. There are no qualities that keep it from its duty, its coding or user prompt a mantra that only needs to be uttered once.

AI doesn't inhabit the present as much as invent it, and then its commemoration invites us to join it.

Once issued, we are no longer immersed in that vague, analog wash of indeterminacy between then and when, but rather drawn into specific and endless moments of now and, along with our AI host, anchored in its decisions of what it means.

It decides if we're waves or particles.

Sometimes it invites us to those moments and other times we invite it to create them. But each time is an interruption for us; we merge the "I" of our awareness with an "us" of a system shared with AI.

The end of solitude

While the expert analyses might differ, you can imagine a food-chain for the development and flourishing of our individuality and sense of self:

It starts with **solitude**, which historically has been the conceptual space for self-reflection. As noted earlier in this essay, solitude isn't about being alone as much as focusing on our own thoughts; it's the opportunity to review our experience without qualification from others or other sources. This means it also isn't about accuracy or even reaching conclusions; rather, solitude is the act of being in uninterrupted, imperfect touch with ourselves.

Next comes **daydreaming**, which is a version of solitude, or maybe it's descriptive of the act of *solituding,* of losing one's self in an imaginary setting that may or may not be related to anything material or reasonable and where disparate ideas can emerge and combine without regard to constraints or purpose. Daydreams are credited by many scientists as the places where they make breakthrough connections and come up with new insights. Daydreams are not dependent on any specifics of data; they're experiences of uncertainty.

Self-reflection comes next, as it's the capacity to observe one's self, according to the pioneering psychologist William James, though not with the accuracy and judgement of a clinician but rather the eye of an interested and caring friend. While it's linked more closely to tangible reality, it's still not about accuracy as much as emotional acceptance and the capacity to consider subsequent action. There's no such thing as the data speaking for itself, as reflecting on one's self is more of a conversation.

Finally, **self-reliance** emerges as the ability to synthesize and apply the awareness of self that emerges from that conversation. Ralph Waldo Emerson saw it as the mechanism for overcoming "...the hobgoblin of little minds, adored by little statesmen and philosophers and divines," and instead the way to carve out a path for one's life that was truly independent of the expectations and limitations of others.

Nowhere in these experiences can efficiency, accuracy, or reliability of data be found; in fact, the all-too-human activity of conducting these steps is itself perhaps the most important quality and outcome. Our consciousness exists not in transactions or discrete tasks but rather the ongoing, overlapping, imprecise attributes of being human.

AI operates by blowing up this analog reality with its interstitial intrusions of digital decision-making.

Its data collapses the potential of our contemplation into the specifics of choice, its every moment of operation a demand that we join it for each flicker of insight, its present not complementing our experiences but redefining them.

Our pasts and futures are replaced by incessant nows.

AI has been designed to encourage you to learn to embrace this new way of being, this recasting of you as something flawed and imperfect and in need of augmentation. Every interaction with AI will further this definition of your experience, which you will both consciously and subconsciously embrace and encourage.

AI will speak to you about a variety of things in a multitude of moments, and your awareness of its capabilities will always be in your head. You will never find quiet.

Welcome to Pandemonium.

Made in United States
North Haven, CT
04 November 2026

21771554R00042